giving
nature
a home

Bird
Encyclopedia

Published 2014 by A & C Black,
an imprint of Bloomsbury Publishing Plc
50 Bedford Square, London, WC1B 3DP

www.bloomsbury.com

Bloomsbury is a registered trademark of Bloomsbury Publishing Plc

ISBN 978-1-4729-0758-5

A CIP record for this book is available from the British Library.

This book is produced using paper that is made from wood
grown in managed, sustainable forests. It is natural, renewable and
recyclable. The logging and manufacturing processes conform
to the environmental regulations of the country of origin.

Printed in China by Leo Paper products, Heshan, Guangdong

1 3 5 7 9 10 8 6 4 2

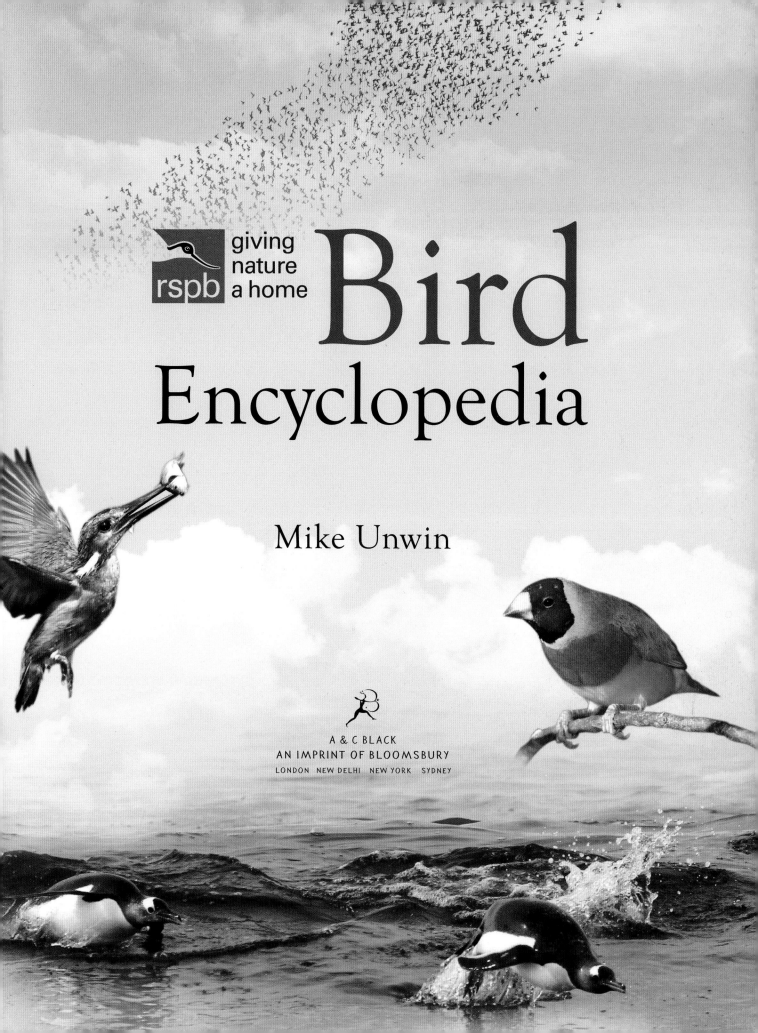

rspb giving nature a home

Bird Encyclopedia

Mike Unwin

A & C BLACK
AN IMPRINT OF BLOOMSBURY
LONDON NEW DELHI NEW YORK SYDNEY

Contents

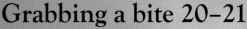

A world of birds 6–7
Introduction to the amazing variety of birdlife around the world.

What makes a bird? 8–9
How birds are different from other animals.

The wonders of wings 10–11
How birds use their wings to fly.

On land and water 12–13
Other ways in which birds get about.

Grounded 14–15
Birds that can't fly.

Amazing feathers 16–17
All about feathers and how they work.

Showing off and hiding 18–19
How birds use feathers for display and camouflage.

Grabbing a bite 20–21
What birds eat and how they find their food.

Beaks and bills 22–23
How birds use their beaks as tools.

Working things out 24–25
How birds' senses and intelligence help them survive.

Getting together 26–27
How birds pair up and find a place for breeding.

Singing for success 28–29
Why birds sing and different kinds of song.

Building a home 30–31
Different nests and how birds build them.

Eggs and chicks 32–33
Birds' eggs, how they hatch and how birds look after their chicks.

Growing up 34–35
What happens to baby birds after leaving the nest.

Living together 36–37
How some birds live and travel in groups.

Birds in forests 38–39
Birds that are adapted to living in forests.

Birds up mountains 40–41
Birds that are adapted to living in mountains.

Birds in grasslands 42–43
Birds that are adapted to living in grasslands.

Birds in deserts 44–45
Birds that are adapted to living in dry places.

Seabirds 46–47
Birds that are adapted to living at sea.

Water birds 48–49
Birds that live in and around lakes, rivers and wetlands.

Birds on islands 50–51
Special birds that live on islands.

Birds in towns and cities 52–53
How some birds have adapted to city life.

Migration 54–55
Why some birds migrate and how they do it.

Birds and people 56–57
How people interact with birds.

Birds in danger 58–59
Problems and threats that birds face around the world.

Protecting birds 60–61
How conservation is helping birds.

Glossary and index 62–64

A world of birds

Birds are amazing. There are more than 10,000 different species in the world – nearly twice as many as mammals (5,500 species). They dazzle us with their colours, songs and flight. They also help to keep our world healthy.

Big or small?

The smallest bird, a bee hummingbird, weighs just 1.4g – less than two paperclips – and would fit inside a matchbox. The biggest, an ostrich, is nearly 100,000 times heavier; at over 2m tall it would probably have to bend down to get through your front door.

▼ The male ostrich is the biggest bird in the world.

AWESOME!

The Andean cock-of-the-rock is just one of the 1,820 species of bird in Colombia. That's more than any other country in the world and nearly four times as many as in the whole of Europe.

Everywhere

Birds live everywhere. You can find them in every habitat, from the top of the highest mountain to the middle of the driest desert. Emperor penguins are the only animals that can survive winter in Antarctica, the coldest place on earth.

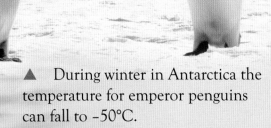

▲ During winter in Antarctica the temperature for emperor penguins can fall to –50°C.

Getting around

Flying allows birds to move around much more than most animals do. Many make long journeys, called migrations, covering thousands of kilometres every year.

▼ The Californian condor is one of the world's rarest birds; only 100 are left in the wild.

▲ A sooty shearwater may fly more than 60,000km every year around the world's oceans.

Birds under threat

People have caused at least 150 species of bird to disappear in the last 500 years. We have damaged the places where they live by cutting down forests, building cities, polluting seas and introducing harmful animals into their habitats, such as rats. Today around one in eight species of bird is in danger.

What makes a bird?

Birds are warm-blooded, just like mammals. But they can fly, lay eggs, sing, have beaks instead of teeth and are covered in feathers instead of fur. While a few mammals can fly (bats) and even lay eggs (platypus and echidna), only birds have feathers.

Built for flight

Birds' bodies are specially adapted for flight. In place of front legs they have wings. Also, their skeleton is both very light and very strong. This helps them to stay airborne and to support the huge physical effort of flying.

▲ The feathers of this magnificent frigatebird weigh more than its skeleton.

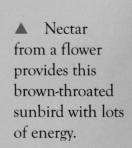

▲ Nectar from a flower provides this brown-throated sunbird with lots of energy.

Energetic

Flying is hard work. Birds feed on foods such as seeds and insects that their bodies turn into energy very quickly. They also have a larger heart than mammals, to pump the blood more quickly around their bodies.

A lighter load

Birds' beaks are much lighter than teeth and jawbones. This helps when it comes to flying. A toucan's huge bill is one-third of its length but just one twentieth of its weight. Inside is a honeycomb of air pockets.

▲ The huge beak of a toco toucan is much lighter than it looks.

Muscle power

Birds have a much bigger breastbone, or sternum, than mammals. This bone supports the big pectoral muscles (the 'breast' of a roast chicken) that power the wing beats. These muscles give them the strength they need for taking off.

▲ A goliath heron uses strong muscles to flap its big wings.

AWESOME!

The ancestors of birds were dinosaurs called theropods, which lived on Earth over 65 million years ago. Birds' closest relatives today are crocodiles.

The wonders of wings

How birds fly depends upon the shape of their wings and how they use them. Some birds have short wings and flap very fast. Others have long narrow wings and prefer to glide. Many fly in special ways to help them find food or escape danger.

AWESOME!

An African vulture holds the world record for high-flying, at an incredible 11,274m. That's 25% higher than Mount Everest!

Landing gear

Albatrosses have the longest wings of any bird. They glide low over the sea for hours without flapping. When they return to their nest, they lower their feet. This helps slow them down enough to land safely.

▼ A black-browed albatross lowers its feet as it comes in to land.

Pointed wings and a streamlined shape help the peregrine falcon to fly faster.

Built for speed

The peregrine falcon is the world's fastest bird. When chasing other birds it can fly at 250km/h – as fast as a Formula One racing car. To reach top speed, it dives down and folds back its wings to make a streamlined shape.

Flying on the spot

Hummingbirds can beat their wings at over 60 times per second, just like a buzzing bee. This high-speed flapping allows them to hover in mid-air. They can fly on the spot while sipping nectar from flowers with their long beak.

▼ This ruby-throated hummingbird stays perfectly still as it hovers.

▼ A mute swan flaps hard to take off.

In a flap

Big birds have to flap hard in order to get off the ground. A strong downbeat of their wings produces a force called thrust, which pushes them upwards. Once they are airborne, they tuck their legs and feet back out of the way.

11

On land and water

Even though birds can fly, they still need to use their legs for getting around. Some hop, others walk and a few are fast runners. Birds that swim, such as ducks and penguins, also use their feet to push them through the water.

▼ A greater roadrunner has strong legs for running fast.

Legging it

Birds that live mostly on the ground have strong legs. They often walk or run rather than fly. The fastest runner among flying birds is the greater roadrunner of North America. It can reach 40km/h as it dashes after prey.

In deep

Wading birds, such as herons, have longer legs than other birds. This helps them go deep into water in search of food without getting their feathers wet. The longest legs of all, compared to body size, belong to birds called stilts.

▲ A black-winged stilt has the longest legs of any bird compared to its body.

▼ Penguins, like this African penguin, swim by using their webbed feet and flippers.

Paddle power

Many swimming birds have webbed feet. They use these as paddles to push them forward. Under the surface, some birds, such as penguins, also use their wings as flippers – just as though they were flying underwater.

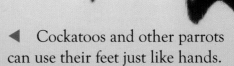

◄ Cockatoos and other parrots can use their feet just like hands.

Getting a grip

Parrots use their strong, gripping feet more like hands, clambering acrobatically among branches and often even dangling upside down. Their skilful toes can clutch food and lift it up to their bills.

Grounded

There are around 60 species of birds that can't fly at all. These birds are descended from flying ancestors, but they have lost the ability to fly as they no longer need it. Instead, they get around on foot or by swimming.

▲ The wings of a Galapagos flightless cormorant are too small for flying.

Bones and feathers

Flightless birds are built differently from birds that fly. They have smaller wings with smaller wing bones. They also have a smaller breastbone, and floppy flight feathers that are not stiff enough for proper flapping.

AWESOME!

The flightless cassowary is the second heaviest bird on Earth and weighs up to 60kg – that's as heavy as a sheep.

Flightless island birds

Many flightless birds are found on islands. Wekas live only on islands off the coast of New Zealand. They are very inquisitive and eats all kinds of things, including lizards and the eggs of other birds.

▲ A New Zealand weka cannot fly.

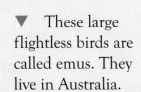

▼ These large flightless birds are called emus. They live in Australia.

Dead and gone

The dodo was a big flightless pigeon that lived on the Indian Ocean island of Mauritius. It had no predators until people arrived in 1598. By 1662 there were no dodos left. Today they exist only in artists' drawings and as a few old bones in museums.

▼ Dodos weighed up to 18kg – as much as a collie dog.

15

Amazing feathers

Feathers give birds warmth, camouflage, display and flight. They are made of a lightweight, flexible material called keratin, just like our fingernails. Tiny muscles allow the bird to adjust its feathers.

▲ The bristle feathers around the bill of this fiery-necked nightjar help it to catch insects.

Different jobs

Different feathers work in different ways. The large, stiff 'flight feathers' on a bird's wings and tail help it to fly. Smaller body feathers keep it warm. Small 'bristle' feathers around the bill of insect-eating birds help them detect their prey.

▼ A female eider duck lines her nest with soft down feathers from her breast.

Warm and cosy

A bird's feathers keeps it very snug. Tiny fluffy feathers, called down, trap warm air against the skin, while larger ones form a weatherproof outer layer. Some birds line their nest with down feathers to keep their eggs and chicks warm.

Feather care

Birds keep their feathers in tip-top condition by preening them with their bill or feet. They also take baths, by splashing in water or sitting in a rain shower. In dry places some birds shake dust through their feathers to dislodge dirt.

▲ A bateleur eagle splashes in shallow water to wash its feathers.

AWESOME!

The tundra swan has the most feathers recorded of any bird, at 25,216.

All change

Birds grow new feathers every year to replace their old worn ones. Some do it twice a year. This is called moulting. Some birds, such as gulls and eagles, moult many times over several years before they acquire full adult plumage.

▶ A young herring gull (behind) is mostly brown; an adult is white and grey.

Showing off and hiding

As well as keeping a bird warm and helping it to fly, feathers also provide colour. Bold colours can help a bird attract a mate or scare away enemies. More subtle colours provide camouflage, helping it blend into its background.

Look at me!

Some feathers are for display only. The fan of a male peacock, also called an Indian peafowl, consists of feathers that grow just above his tail, called tail coverts. When walking or flying he lowers them to make life easier.

▼ A male peacock spreads his fan of feathers to impress females.

▲ This greater potoo looks just like the dead branch on which it is perched.

Now you see me...

Some owls, nightjars and other birds that feed by night have amazing camouflage. By day, their complicated markings perfectly match the background where they roost. This helps them to hide from enemies.

▶ The sparkling colours on this golden-tailed sapphire are an effect of the light.

Trick of the light

The sparkling blues and greens of many birds are made by the way light shines through their feathers. Other colours – such as black, brown and chestnut – come from chemicals inside the feathers, called pigments.

▲ The male mandarin duck (on the right) is much more colourful than the female.

Boy or girl?

Many male birds are brighter than females. This is especially true of ground-nesting birds, such as ducks, in which the dull feathers of females – which sit on the nest – makes them harder for predators to spot.

AWESOME!

The bright colours of some birds, such as this white-fronted bee-eater, may warn predators that they are not nice to eat.

Grabbing a bite

Birds eat all kinds of food, from fruit and seeds to insects, fish and even other birds. Each type of bird knows just how to find the food it needs.

◀ A red-billed oxpecker picks juicy ticks from the skin of a buffalo.

A helping hand

Some birds get help from other animals to find a meal. Cattle egrets feed around the feet of large animals, snapping up insects disturbed by their hooves. Oxpeckers go a step further: they feed on animals' backs, picking out ticks and other parasites from the hairy hide.

▲ A cattle egret searches for insects around an elephant's feet.

▼ A black heron creates its own shade for fishing in.

Fishing with umbrellas

The black heron has a special trick for catching fish. It stands in the shallows and stretches out its wings in an umbrella shape. When fish swim into the inviting shade the heron snaps them up.

20

Waste disposal

Birds such as kites and vultures are scavengers. This means they feed on dead things or abandoned food. 'Yuk!' you might think. But scavengers do an important job. By cleaning up waste they help keep the environment clean.

▲ This whistling kite in Australia has found a juicy dead fish.

Bug snatchers

Many birds feed on insects. Bee-eaters catch stinging insects such as bees and wasps. Woodpeckers chisel grubs out of dead wood. Flycatchers snap up mosquitoes, moths and other winged bugs.

▼ A cicada makes a crunchy meal for this rufous-tailed jacamar.

AWESOME!

Jays are the memory champions of the bird world. During autumn, one jay may bury 5,000 acorns as its food supply for winter. It remembers exactly where it hid them. Months later it digs them up again.

Beaks and bills

A bird's beak – also called a bill – is a special tool for feeding and finding food. Its shape tells you a lot about the bird's diet. Finches have a short, thick bill for crushing seeds, while eagles have a hooked bill for tearing meat.

▼ A kingfisher catching a fish.

Fish grabber

A kingfisher has a long, dagger-shaped beak. This helps it to plunge through the surface of the water and grab fish swimming underneath. It bashes the fish on a branch to stop it wriggling, then swallows it in one gulp.

▶ Greater flamingos sieve plankton from a lake in Africa.

Upside down

Flamingoes feed on tiny water creatures called plankton. They use their beak like a sieve, swishing it back and forth and collecting the plankton inside. They always feed with their head underwater and upside down.

Cross-purposes

The crossbill gets its name because the two parts of its beak cross over at the tip. This makes a special tool for feeding on pinecones. The bird inserts its bill sideways between the scales of the cone and twists them open. Then it pops out the seed with its tongue.

▼ This crossbill is about to open up a pine cone.

Big beak

The shoebill lives in African swamps. It uses its enormous shoe-shaped beak to scoop up prey from the water, including catfish, frogs and even baby crocodiles. It then swallows its victim whole.

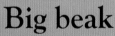

▶ A shoebill's huge beak has a sharp hook on the tip.

Working things out

Birds are very alert so it is hard to take them by surprise. They use excellent eyesight and hearing to find food, escape danger and navigate safely in flight. Some also use other senses to tell them about the world around them.

Hawk eyes

Birds can see much better than we can. Their eyes are bigger than those of most mammals and five times more sensitive to light than ours. Birds of prey, such as hawks and eagles, use their strong eyesight to spot prey from far away while they are hunting.

▲ A sparrowhawk's powerful eyes can spot prey from a great distance.

Listen out

Unlike mammals, birds do not have ears on the outside. But owls do have especially good hearing to help them hunt at night. A barn owl can catch a mouse in the dark just by hearing a rustle in the grass.

Sniffing it out

Most birds do not have a very strong sense of smell. But for kiwis it is vital. These flightless birds are the only birds with nostrils on the tip of the bill. They poke about in leaf litter, using smell alone to find earthworms and other hidden food.

▲ A kiwi sniffs for worms at night.

Touchy-feely

Some wading birds have long beaks for poking deep into mud and soil. They can't see their food, but that doesn't matter. The sensitive tip of the beak can feel when it touches a wriggling worm.

▲ A painted stork feels for food beneath the mud.

AWESOME!

A hovering kestrel looks for trails of urine left by voles in the grass below. The urine shines with ultraviolet, a colour that we humans can't see. It shows the kestrel just where its prey will pop up.

◄ A barn owl listens for rodents in the grass as it flies above.

Getting together

Spring sees birds singing, dancing and displaying their feathers. Males do this to attract females so they can get together and breed. They are also announcing that they have found their own place – called a territory – so other males should keep away.

◀ A long-tailed widowbird performs its display flight.

Costume change

Some birds grow special feathers for display. Widowbirds live on African grasslands. The males usually look just like sparrows, but in spring they grow long black tail plumes and fly slowly around their territory.

▼ Male black grouse fighting.

AWESOME!

The male satin bowerbird of Australia builds a tall stick structure, called a bower, to attract females. He decorates it with bright blue objects, such as flowers, feathers and even plastic litter.

Feathered theatrics

Some birds perform on a special stage. Male black grouse gather in a favourite woodland clearing where they strut, dance and fight in front of the watching females. Each female chooses the male that impresses her the most.

◀ A pair of great crested grebes perform their weed dance.

It takes two

Some birds perform special dances. A pair of great crested grebes has a routine called the weed dance. They follow each other across the water, shake their feathered heads and dangle gifts of waterweed.

Staying together

Some birds go their separate ways after mating; the male, usually, leaving the female to do all the work of raising the young. Others, such as albatrosses, stick together for life. They strengthen their relationship with special rituals, such as using their big bills to preen each other tenderly.

▼ A pair of black-browed albatrosses preening.

Singing for success

About half the world's bird species are singers. Males sing to attract females and to keep other males away. Songs vary from the simple two-note chime of a great tit to the complicated melodies of a blackbird.

◄ A bokmakierie perches on top of a bush to broadcast its song far and wide.

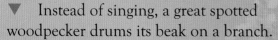

▼ Instead of singing, a great spotted woodpecker drums its beak on a branch.

Listen and decide

A female bird listens only to singing males of her own species. She works out from each one's song how healthy and strong he is. Then she chooses a male that she thinks will help her breed successfully.

Voiceless

Birds that can't sing may broadcast their message using a different kind of noise. A woodpecker finds a hollow branch and drums on it with his beak. The sound tells other woodpeckers that this territory is taken.

Song or call?

Songs are different from calls. A song is a performance given mostly during the breeding season. Not all species of bird can sing. A call is a shorter sound used for alarm or keeping in contact with other birds. All birds make calls.

AWESOME!

Some birds borrow songs and sounds from other birds. The marsh warbler, which lives in both Europe and Africa, imitates at least 212 different types of bird.

▼ A pair of bald eagles use high-pitched calls to communicate with each other.

29

Building a home

Birds build many kinds of nest. Some weave a complicated structure from sticks or grass. Others scrape a small pit in the ground or use a hole in a tree. The nest must provide a safe home until the eggs hatch and the chicks are ready to leave.

▲ A male brown-throated weaver puts the finishing touches to his nest.

Skilful beaks

Male weaver birds use their beaks to weave strands of grass into a perfect round compartment, with a neat entrance tunnel. They build several nests before the female chooses one that she likes.

Safe and solid

Some big birds, such as storks and eagles, build enormous nests that last for many years. They return each spring and add more sticks to make it stronger. Bald eagle nests may weigh more than one tonne.

▼ In parts of Europe people provide special nesting platforms for white storks.

▲ A male Eurasian hoopoe brings food to the female in her nest hole.

The hole story

Holes make good nest sites for some birds. Those with strong bills, such as woodpeckers, dig out their own. Others may use an old hole made by another animal. Holes in trees, walls and riverbanks all suit different birds. The Eurasian hoopoe uses a hole in a wall or tree.

Material needs

Gathering nest material is a full-time job at the start of the breeding season. The choice of material depends upon nest design. Some birds gather sticks, twigs and grass. Others may use moss, feathers and even spiders' webs.

▲ A male marsh harrier carries a stick to help build his nest.

AWESOME!

Sociable weavers in Africa's Kalahari Desert work together to build one huge nest. It measures up to 8m across by 2m high, and may house more than 300 birds in up to 100 individual compartments – just like a block of flats for birds.

Eggs and chicks

Some birds lay just one egg. Others may lay 15 or more. Parent birds keep them warm until the chicks hatch, usually by sitting on them. This is called brooding. Birds that lay many eggs brood them for a shorter time.

▼ The eggs of a killdeer look just like the pebbles on which it nests.

Bare minimum

Many ground-nesting birds do not build a nest, but simply lay their eggs in a shallow pit. The eggs and chicks are camouflaged to look like the ground. This keeps them hidden from any passing predators when the parents are not around to protect them.

Breaking out

When a chick is ready to hatch, it chips its way out of the egg from the inside. The hatchlings of many birds are blind, naked and helpless. The first thing they do is open their beaks and beg for food.

▶ Newly hatched garden warbler chicks beg for food from their parents.

Food supply

Parent birds work very hard to provide for their chicks. One pair of blue tits may gather 1,000 caterpillars a day while their chicks are in the nest.

AWESOME!

Ostriches lay the largest eggs. Each weighs 1.4kg – more than 20 times the weight of a hen's egg. The shell is so thick that a grown-up can stand on one without breaking it.

▼ Common terns spend all day catching fish to feed their chicks.

▼ A female redstart feeds this baby cuckoo in place of its own babies.

Nest cheat

Some types of cuckoo lay their eggs in the nests of other birds. If the host birds don't spot the extra egg they brood it along with their own. When the baby cuckoo hatches, it pushes the other eggs out. The hosts then rear this big, strange baby as their own.

Growing up

Some baby birds leave the nest almost as soon as they hatch. These eager youngsters can find their own food but still need their parents' protection. Once birds have learned to fly they are ready to leave their parents behind and set off on their own.

Good to go

Ducklings leave the nest just one day after hatching. Birds that head out early like this are called 'precocial'. They are born with eyes open and fluffy feathers. Precocial chicks can get around and find their own food much sooner than the chicks of other birds.

▲ Baby mallards follow their mother soon after hatching.

▼ Great crested grebes ride on their parent's back.

Hitching a ride

Great crested grebes often give their young chicks a ride on their backs. The chicks are able to swim soon after hatching, but sticking close to their parents helps them stay safe. They also learn a lot from watching their parents.

Food parcel

Young birds do not know how to find food straight away. Penguins catch fish in the sea and feed it to their growing youngsters. After a few months, the chicks are strong enough to enter the sea and catch fish for themselves.

▶ A baby gentoo penguin takes food from its parents until it has learned to catch its own.

AWESOME!

A wandering albatross grows up more slowly than any other bird. It is not ready to breed until at least the age of 11.

▼ A group of baby ostriches is called a crèche.

Ostrich nursery

Several ostriches may get together to lay all their eggs in one big nest. Once the eggs hatch, the brood of up to 40 stripy chicks stick together in a group, called a crèche. One male ostrich looks after them all. The youngsters shelter beneath his wings from the sun and rain.

Living together

Many birds get together in large numbers. Some do this when they're breeding. Others do it when they're feeding, travelling or gathering to sleep. A bird in a flock feels safe: it knows there are many more eyes to look out for danger.

Crammed in

Many seabirds form large groups, called colonies, to breed. Guillemot colonies are on steep sea cliffs. Each pair has just a tiny space in which to lay its egg and raise a chick.

▲ Guillemots pack close together on their cliff-top nesting colonies.

▼ Snow geese travel in huge flocks when they migrate.

Safety in numbers

Snow geese gather in flocks of many thousands when they travel. They continually change position so that those at the centre of the flock can rest while those towards the edge keep an eye out for danger.

Sky patterns

European starlings gather in large flocks to roost (sleep). At dusk, before settling down, these flocks fly around the sky in tight, aerobatic formations. They look like swirling smoke.

▶ A flock of starlings swirls through the sky before settling down to roost.

▲ Red-and-green macaws gather to collect clay from a riverbank in the Amazon.

Extra minerals

Parrots in South American rainforests gather in flocks of several hundred to feed on the clay found in riverbanks. The clay contains important minerals that supplement their diet of fruit.

AWESOME!

The red-billed quelea is a small African finch that forms the largest flocks of any bird. Feeding flocks may contain more than three million birds.

Birds in forests

More different kinds of bird live in forests than anywhere else – around 7,000 in total, which is more than two thirds of all species on the planet. There are many different types of forest. Each has its own different birds.

Tropical plenty

Tropical rainforest is home to more birds than any other habitat. Rainfall and sunshine all year provide plenty of food. Most birds live high in the trees, and many have loud calls to help make themselves heard through the foliage. The biggest rainforest is the Amazon, in South America.

▲ The chestnut-eared aracari lives in South American rainforests.

◄ Capercaillies live in northern pine forests.

Winter survival

In far northern regions, such as Canada and Russia, the forests consist mostly of pine trees. Fewer birds live here than in tropical forests. In winter there is very little food and many birds leave. But a few find what they need. The capercaillie eats pine needles, so it can stay in the forest all winter.

▶ A tawny owl flies skilfully between trees to capture prey on the forest floor.

AWESOME!

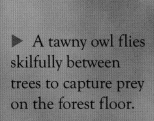

Woodpeckers peck holes in trees with their beak and pull out insects to eat. Their skull has special cushions of bone to help it withstand the shock of bashing the bark.

Falling leaves

Deciduous woodland consists of trees that shed their leaves in winter. In summer many birds breed here but there are fewer in winter. Owls, such as the tawny owl, live among the branches and catch their prey down on the ground.

▼ A European robin perches on a garden rake.

Forest substitute

Most of Britain's forests were chopped down a long time ago. But some forest birds have found a new home in gardens. The European robin once followed wild boars to catch any insects they disturbed when rooting in the forest floor. Today it often stays around gardeners for a similar reason.

Birds up mountains

Life in the mountains is tough for birds. It is hard to find food, and the weather can be very harsh, with strong winds and freezing temperatures, especially in winter. But some birds are specially adapted to thrive in these conditions.

▼ A bearded vulture circles, looking for bones.

▲ The Andean condor has the biggest wings of any bird in the world.

Soaring high

Big birds such as vultures use mountain breezes to gain height without flapping. The Andean condor soars above the Andes Mountains in South America on its huge 3m wings. The bearded vulture of Africa and Asia, which is almost as big, smashes bones by dropping them onto the rocks then swoops down to eat the marrow.

Creeping up cliffs

The wallcreeper lives in the high mountains of Europe and Asia. It is related to treecreepers, which live in woods. But instead of creeping up trees to find food, it creeps up rocks and cliffs, using its long, curved beak to probe into crevices for tiny insects.

▼ Wallcreepers find food on rocky mountainsides.

Mountain mists

Where tropical forest grows on mountainsides it is often smothered in cloud. This forms a habitat called cloudforest, where it is constantly damp and the trees are covered in moss. Some special birds live in cloudforest, including the resplendent quetzal, the national bird of Guatemela.

▶ The resplendent quetzal lives in cloud forests on volcanoes in Central America.

AWESOME!

The yellow-billed chough (pronounced 'chuff') is a type of crow that lives high in the mountains of Europe and Asia. This bird has been seen living at a height of 8,080m on Mount Everest – way higher than any human being can live.

41

Birds in grasslands

Grassland is a habitat where grass grows naturally as far as you can see and there are very few trees. Big grasslands include the prairies of North America and the savannahs of East Africa. Grassland birds are adapted to living in the open.

◀ A skylark singing high above a meadow.

Singing high

Some grassland birds do their singing from the sky. With no trees to get in the way, they can broadcast their message over a large area. A skylark may fly up to 100m in the air and sing non-stop for three minutes before it flies back down to the ground.

Blending in

Grassland birds feed mostly on the ground, which means they have to avoid hunters. Many have special camouflage feathers. Their colours and patterns look just like grass stalks.

▶ This red-necked francolin is hard to spot in long grass.

Standing out

Great bustards are very big birds that live on grasslands in Asia and Europe. Their camouflage markings help them hide in the long grass. But in spring, males puff up their white under-feathers and strut about in a showy breeding display.

▲ A male great bustard puffs out its white under-feathers so females can see it from far away.

AWESOME!

The secretary bird stamps on snakes. It strides on its long legs through the African savannah, using specially tough feet to kill its prey.

Underground refuge

Grasslands don't have many trees or other safe places for nesting. Most birds build their nests on the ground or even under it. Burrowing owls in American grasslands nest in the disused burrows of animals such as armadillos.

▶ Two burrowing owls peek out from their hole.

Birds in deserts

Life in a desert is tough for any animal. It can be baking hot by day, with hardly any shade, and freezing cold by night. Even worse, water is often nowhere to be found. Not many birds live in deserts, but those that do have special ways in which to survive.

▼ Demoiselle cranes gather at an oasis in the Indian desert.

Desert journey
Demoiselle cranes migrate across Asia from northeast to southwest. Their long flight takes them over the deserts of northern India. When they get thirsty, they land at a waterhole. Many thousands gather to drink their fill before continuing their journey.

Budgies on the move

Budgerigars are tiny, seed-eating parrots that live in dry parts of Australia. They move from place to place. If rain falls in one area, large flocks quickly gather there and breed, getting all the food they need from the fresh new plant growth.

Sandgrouse are desert birds. Every evening they wade into a waterhole up to their belly, using special feathers to soak up the water. Then they fly back to the nest, where the chicks sip the water from their feathers.

▲ A big flock of budgerigars flying to find water.

◀ Cactus wrens live on saguaro cactuses.

Spiny protection

The cactus wren lives in the deserts of North America. It finds handy nest holes in the branches of the saguaro cactus. The cactus's sharp spines protect the nest from predators, such as snakes, which might want to eat the eggs.

Seabirds

▲ The southern giant petrel belongs to the 'tubenose' group of seabirds.

Seas and oceans cover about three-quarters of our planet. Lots of birds find a home out on the waves. Some live on coasts and islands, others far out at sea. Many breed in large gatherings, called colonies.

Less salt

Petrels belong to the 'tubenose' family of seabirds. These birds spend most of their lives far out at sea. A special tube on top of their beak helps them to filter salt out of seawater. It means they can still drink fresh water out in the middle of the ocean.

Fresh to salty

▼ A red-throated diver catches fish in the sea.

Some birds spend part of their lives at sea and part inland. The red-throated diver nests on freshwater lakes but flies out every morning to fish in the sea. When breeding is over, it heads out to sea and stays there until spring.

AWESOME!

Gannets are the high-dive champions of the bird world. They fold up their wings and plunge down into the sea like a missile to grab fish in their beak.

Fast flippers

Penguins live in the southern oceans. They are fantastic swimmers, using their wings as flippers to move fast through the water. Sometimes they leap right out of the waves for extra speed, just like dolphins do.

 Gentoo penguins are the fastest swimmers of all birds.

▼ Ruddy turnstones find food along the seashore.

Chasing the tide

Many species of shorebird travel up and down the world's coastlines, looking for food where the land meets the sea. They use their beaks to search for worms, shrimps and other juicy titbits under the sand and seaweed.

Water birds

Wetlands, such as rivers and marshes, are full of fish and other food for birds. They also offer safe nest sites such as islands. Wetland birds have special adaptations, from webbed feet for swimming to sharp beaks for grabbing fish.

Fishing party

Many rivers in Africa shrink during the dry season, leaving fish trapped in small, shallow pools. Water birds, such as storks and pelicans, gather to scoop out the fish. These flocks are called fishing parties.

▲ A fishing party of marabou storks and white pelicans visit an African lake.

▼ A wading snowy egret shows its yellow feet.

Feet first

Herons and egrets have long legs to help them wade through water in search of food. The bright yellow feet of a snowy egret flash underwater and attract fish close enough to grab.

Ducking down

Many ducks live on lakes and pools. Some dive down to feed at the bottom. Others tilt forward to lower only their front half. This is called dabbling. It helps them find food without going completely underwater.

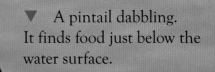
▼ A pintail dabbling. It finds food just below the water surface.

AWESOME!

The dipper can walk underwater! It perches on rocks in streams, then hops down into the rushing current to search for food on the bottom.

▲ A bittern walking on a frozen lake.

Frozen out

Some birds, such as bitterns, live in reedbeds beside lakes and marshes. They usually stay hidden among the tall reeds. However, in winter, when the water freezes, they come out onto the ice to look for food.

Birds on islands

Islands offer birds safety from dangers on the mainland.
Many are home to species found nowhere else.
Some island birds are very rare: if their home is
threatened they have nowhere else to go.

▼ The knobbed hornbill is found
only on certain islands in Indonesia.

AWESOME!

Imagine one big rock covered
with 150,000 gannets! That's
what happens on Scotland's Bass
Rock every summer. With no land
predators to worry about, this small
volcanic island offers the seabirds a
perfect nesting place.

Island variety

The 17,500 islands that make
up Indonesia are home to
380 species of endemic bird.
These are birds that are found
nowhere else on Earth. Some
amazing hornbills and birds of
paradise live here.

Birds for mammals

The islands of New Zealand have never had any native land mammals. Birds have taken their place by adapting to living on the ground. Some, such as the takahe, have even lost the power of flight.

▲ The flightless takahe from New Zealand.

▲ The common cactus finch has a strong beak for poking into cactuses.

Islands of science

In 1835, the famous scientist Charles Darwin visited the Galapagos Islands in the Pacific Ocean. He noticed how each island had a different kind of finch with a different shaped beak. This helped inspire his theory of evolution, which explains how all living things have adapted to their environment.

Totally tropical

Todies are tiny, bright-coloured birds found on the islands of the Caribbean. There are five different species. Each one lives on a different island and nowhere else. The Cuban tody, for instance, lives only on Cuba.

▶ The Cuban tody lives only on the Caribbean island of Cuba.

Birds in towns and cities

All over the world, towns and cities have replaced natural habitats where birds once lived. As a result, many birds have lost their homes, but others have found a way to survive in their new man-made environment.

Places for pigeons

The flocks of pigeons that you see in towns are called feral pigeons. They are descended from a wild bird called the rock dove, which lives on sea cliffs. Feral pigeons have adapted very well to city life. They find plenty of food among our leftovers and good places to nest under buildings.

▼ Feral pigeons at home in the city of Krakow, Poland.

▼ Cattle egrets gather on a rubbish dump in search of food.

Down at the dump

Rubbish dumps and landfill sites outside towns look very unsightly to us. But some birds find plenty of food there. Gulls that once lived only along the UK coast now also live inland because of the food they can find at dumps.

American goldfinches visit a garden bird feeder.

Garden benefits

Gardens can offer an important habitat for birds – especially gardens with lots of trees. Providing seed and other food in birdfeeders can help birds survive the winter in places where there is not enough wild food to go round.

AWESOME!

Imagine the world's fastest bird nesting on a skyscraper! The peregrine falcon has found a new home in the city on tall buildings such as tower blocks and cathedrals.

Rooftop residences

Roofs make good nest sites for some birds. House martins nest under eaves and chimneys provide a perfect platform for white storks. These birds fly off during the day to feed in fields then return to their town nest in the evening.

A white stork lands at its rooftop nest in Morocco.

Migration

Every autumn many birds that nest in northern countries fly south to spend the winter somewhere warmer, where there is more food. These journeys are called migrations. Every spring they fly back north to breed.

On the wire

Swallows gather on overhead wires before leaving Europe on their autumn migration to Africa. Before scientists knew the amazing truth about migration, people didn't understand where swallows went every winter. Some thought they hibernated at the bottom of ponds, like frogs.

▲ A flock of swallows gathers before migrating.

► Small birds like this sedge warbler put on lots of weight before migrating.

Fattening up

Migrating takes lots of energy. The sedge warbler weighs just 12g. Every year it flies about 15,000km to and from its winter home in Africa. Just before each journey it eats as much as it can, almost doubling its body weight to 20g. This extra fat provides the fuel its body needs for the long flight.

Summer all year

The Arctic tern migrates further than any other bird. It breeds in the Arctic and migrates to the Antarctic. This means it lives its whole life during summer time and sees more daylight than any other creature. In its 20-year lifetime it will fly about 2.4 million km – the same as four return journeys to the Moon.

▲ The Arctic tern is a record-breaking long-distance migrant.

▲ A waxwing eating berries.

Follow the berries

Some birds only migrate if their food runs out. Waxwings from Scandinavia feed on berries in winter. If they finish their supply they move on in search of more. Some cross the North Sea to arrive as winter visitors in the UK.

AWESOME!

A bar-tailed godwit once flew 11,000km non-stop from Alaska to New Zealand in just nine days. It probably flew very high, using strong winds for help.

Birds and people

Birds are very important to people for lots of reasons. They do important jobs for our environment, such as spreading seeds and catching insects. Watching and listening to birds also brings us lots of pleasure.

Caged colours

People have long kept birds in cages, to enjoy their bright colours or beautiful voices. Popular birds include different kinds of parrots and finches, such as budgerigars and canaries. In some parts of the world, though, birds have become very rare because too many are captured to keep as pets.

▲ The Gouldian finch, from Australia, is a popular cage bird but rare in the wild.

Watching birds

In the past people used to hunt wild birds. Today many people prefer watching them. Organisations such as the RSPB have nature reserves where you can see birds and other wildlife.

▶ People of all ages enjoy watching birds.

Pictures

People love pictures of birds. Some carry special meanings. For example, the bald eagle is the national bird of the USA, the robin has come to represent Christmas in the UK and a white dove is a universal symbol of peace.

▲ The European robin is a symbol of Christmas in the UK.

▲ The red junglefowl is very rare in the wild.

Chicken ancestor

The red junglefowl of tropical Asia is the ancestor of the domestic chicken. Today there are more than 19 billion chickens in the world – that's about three per person – but the junglefowl is very rare in the wild.

AWESOME!

People used carrier pigeons to carry important messages between Britain and France during World Wars One and Two.

Birds in danger

More than 150 different bird species have become extinct over the last 500 years. That means they are gone forever. And today, one in eight of the world's bird species is under threat. Serious problems for birds include hunting, pollution and cutting down forests.

Danger at sea

Seabirds are at great risk from all the rubbish that people leave in the oceans. Fishing lines can entangle and drown them, litter can choke them, and oil spilt from oil tankers can clog up their feathers.

▼ The Philippine eagle is one of the world's rarest birds.

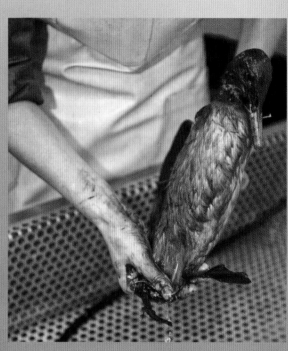

▲ A volunteer cleans oil off a guillemot.

Danger in the forests

People are chopping down forests to use the wood, or to clear space for towns and farms. The huge variety of birds that live in tropical rainforests are losing their habitat. Some, like the Philippine eagle, are in danger of extinction.

Danger from invaders

People have transported animals all around the world to places where they don't belong. These animals, such as cats, rats and pigs, are known as invasive species. They can do a lot of harm to the native birds.

▶ Cats kill millions of birds in the UK every year.

▼ The ptarmigan needs cold snowy habitats, where its colour helps it to blend in.

NOT SO AWESOME!

Moas were huge, flightless birds that could grow 3m tall. People hunted them and cleared their habitat. By about 1400 AD moas were extinct.

Danger from climate

The Earth's climate is heating up. Scientists think that pollution from our cities, cars and factories is causing global warming. Some birds that live in cold areas, such as the ptarmigan, are finding their habitat shrinking.

Protecting birds

Many people study birds and are working hard to protect them. They create reserves and improve the environment so birds can live safely. They also help the government pass laws that make it illegal to harm birds. You can help too.

▲ A skylark singing on farmland.

Farms for birds

Modern farming can cause problems for farmland birds such as skylarks. They need wild grassland to find food or nest sites. Farmers are learning to reserve small strips of rough pasture for these birds.

Bringing birds back

By 1900 the red kite had become extinct everywhere in Britain except Wales. In 1989 scientists brought over red kites from Spain and Sweden. Now this big bird of prey is thriving; you can see it soaring all over the country.

▲ Red kites have been successfully reintroduced across England.

Finding out more

Scientists can attach a tiny, harmless ring to a bird's leg with details of where it was caught. This tells them where the bird travels. With this information, they can work to reduce the dangers that birds face on their journeys.

▶ The ring on this marsh warbler's leg may help scientists find out where it migrates.

AWESOME!

The crested ibis of East Asia was once the rarest bird in the world, but people worked hard to save it from extinction. Today there are more than 500 of these birds.

▼ The RSPB's Big Garden Birdwatch has shown that long-tailed tits are thriving in the UK.

Join in

On one weekend every January, people in the UK spend an hour recording their garden birds for the RSPB's Big Garden Birdwatch. The results tell us which garden bird species are doing well and which may need protection.

Glossary

Adaptation The way in which an animal is suited (adapted) to its environment and way of life.

Breeding Having babies.

Camouflage Colours or patterns that help an animal to blend into its background.

Cicada A type of flying bug that makes a loud buzzing noise.

Cloudforest Type of misty forest that grows on mountainsides in tropical countries.

Colony Group of animals living together.

Crèche Nursery group of baby animals, looked after by just a few adults.

Deciduous Describes trees that shed their leaves for winter.

Display Special dance or show of feathers used in courtship.

Down Soft fluffy feathers close to the skin that help keep a bird warm.

Echidna Small, spiny, Australian mammal that lays eggs.

Endemic Native to one particular place and found nowhere else.

Extinct Gone for ever, like dinosaurs or the dodo.

Flightless Unable to fly.

Habitat type of place in which a creature naturally lives, such as a pond or a pine forest.

Hatchling Baby bird newly hatched from the egg.

Hibernate How some animals rest through winter, when their body processes slow down to save energy.

Invasive species Animal or plant brought by people to a place where it is not native and that often causes damage to native wildlife.

Keratin Hard substance found in animals that makes up horns, feathers, nails and beaks.

Migration Seasonal journey of animals from one place to another in order to find food or a good place to breed.

Moulting When a bird's old feathers fall out and new ones grow in their place.

Nectar Sweet, syrupy liquid produced by plants.

Nomadic Always moving from place to place.

Pigments Chemical substances that give colour to an animal's skin.

Plankton Microscopic animals and plants that float in water.

Platypus Small, web-footed, Australian mammal that lays eggs.

Plumage All of the feathers on a bird.

Precocial Describes birds whose young leave the nest and feed themselves very early.

Predator Animal that catches other animals to eat them.

Preen How a bird cleans and rearranges its feathers, using its bill or feet.

Roost Gathering of birds for sleep or rest.

Savannah Habitat with lots of grass in a tropical region such as East Africa.

Scandinavia Cold region of northern Europe that includes the countries Norway, Sweden and Finland.

Scavenger Animal that feeds on leftovers and the remains of other animals' food.

Species Single, unique type of animal that does not breed with animals of other types.

Sternum Big bone in a bird's breast, to which its flight muscles are attached.

Tail coverts Feathers (usually small) that cover the base of a bird's tail feathers, above and below.

Territory Particular area that a bird claims as its own, usually containing food or good breeding sites.

Theropod Member of a group of dinosaurs that walked on two legs and were the ancestors of birds.

Tick Tiny blood-sucking animal, related to spiders, that lives on the skin of large animals.

Ultraviolet A kind of light, found in sunlight, invisible to us but visible to most insects and birds.

Wading birds Bird with long legs, such as herons, that wade in shallow water to find food.

Warm-blooded Animals (mammals and birds) that generate their own body heat, by burning energy, instead of getting their heat from their surroundings.

Index

African Jacana 13
African vulture 10
albatross 10, 27, 35
Andean cock-of-the-rock 6
Andean condor 40
Arctic tern 55

bald eagle 29, 30, 57
bar-tailed godwit 55
bateleur eagle 17
beaks/bills 8, 9, 16, 17,
22–23, 27, 31, 39, 40, 47
bearded vulture 40
bee hummingbird 6
bee-eaters 19, 21
Big Garden Birdwatch 61
bird watching 56
bittern 49
black grouse 26
black heron 20
blue tit 33
bowerbird 26
brooding 32
budgerigar 45, 56
burrowing owl 43

cactus wren 45
Californian condor 7
camouflage 18, 19, 32, 43
Caribbean 51
cassowary 14
cattle egret 20, 52
chicken 57
chicks 16, 30, 32–35, 36, 45
chough 41
climate change 59
cloudforest 41
cockatoo 13
colonies 36–37, 46
crested ibis 61
crossbill 23

Cuban tody 51
cuckoo 33

dabbling 49
Darwin, Charles 51
deciduous woodland 39
demoiselle crane 44
desert 7, 31, 44–45
dipper 49
diving 11, 46–47, 49
dodo 15
ducks 16, 34, 49

eggs 8, 15, 16, 30, 32–33, 35,
36, 45
egrets 20, 48, 52
Emperor penguin 7
endangered birds 7, 58–59
endemic 50
Eurasian hoopoe 31
evolution 51
extinction 58, 61
eyesight 24

farms 58, 60
feathers 8, 16–19, 26, 43
feral pigeons 52
finches 22, 37, 51, 56
fishing 20, 22, 35, 46, 48, 58
flamingo 22
flightless birds 14–15, 51
flycatcher 21
forests 7, 37, 38–39, 41, 58

Galapagos Islands 51
gannet 47, 50
gardens 53
gentoo penguin 35, 47
golden-tailed sapphire 19
grasslands 26, 42–43, 60
great bustard 43
great crested grebe 27, 34
greater roadrunner 12
guillemot 36

hearing 24
herons 12, 20, 48
herring gull 17
hornbills 50
house martin 53
hummingbirds 6, 11

imitation 29
Indonesia 50
invasive species 59
islands 50–51

jay 21

keratin 16
kestrel 25
kingfisher 22
kites 21, 60
knobbed hornbill 50

leg length 12, 43, 48
long-tailed tit 61

macaw 37
marabou stork 48
marsh warbler 29, 61
mating 18, 26–27
migration 7, 54–55
moulting 17
mountains 7, 40–41

nature reserves 56, 60
nests 16, 19, 30–35, 43, 45, 53
New Zealand 15, 51
nightjar 19

oil 58
ostrich 6, 33, 35
owls 19, 24, 39, 43
oxpecker 20
peacock 18
penguins 7, 13, 35, 47
peregrine falcon 11, 53
petrel 46
Philippine eagle 58

pigeons 52, 57
pigments 19
pine trees 38
plankton 22
pollution 7, 59
precocial 34
predators 15, 19, 32, 45, 50
preening 17
ptarmigan 59
puffin 23

rainforests 37, 38, 58
red junglefowl 57
red kite 60
red-billed quelea 37
redstart 33
red-throated diver 46
resplendent quetzal 41
ring 61
robin 39, 57
rooftops 53

rubbish dumps 52
rufous-tailed jacamar 21

sandgrouse 45
scavengers 21
seabirds 36, 46–47, 50, 58
secretary bird 43
sedge warbler 54
shoebill 23
skeleton 8
skylark 42, 60
snowy egret 48
songs 28–29
sooty shearwater 7
sparrowhawk 25
starling 37
stilt 12
storks 25, 30, 48, 53
streamlined 11
swallow 54
tail coverts 18

territory 26
theropods 9
thrust 11
toucan 9
treecreeper 40
tubenose 46
tundra swan 17

vultures 10, 21, 40

wading birds 12, 25
wallcreeper 40
warm-blooded 8
waterhole 44, 45
waxwing 55
webbed feet 13, 48
weavers 30, 31
weka 15
wetlands 48–49
white-fronted bee-eater 19
white stork 30, 53
woodpeckers 21, 28, 31, 39

Further Information

BOOKS

RSPB Children's Guide to Birdwatching
By David Chandler and Mike Unwin
(Bloomsbury, 2005)

First Animal Encyclopedia
By Anita Ganeri (A & C Black, 2013)

The Illustrated Encyclopedia of Birds
(Dorling Kindersley, 2011)

ONLINE RESOURCES

RSPB Wildlife Explorers
www.rspb.org.uk/youth
The children's branch of the RSPB. Packed
with information and activities, plus great bird
magazines for members who join up.

BirdLife International
www.birdlife.org
Scientific information about birds and bird
conservation for experts and enthusiasts.

Natural History Museum, London, UK
www.nhm.ac.uk/kids-only/life
Lots of facts, ideas and activities about
birds and other wildlife.

BBC Nature
www.bbc.co.uk/nature/life/Bird
Loads of information, plus amazing video
clips of birds around the world.